João Jacinto de Magalhães

Description of a Glass Apparatus

For Making Mineral Waters

João Jacinto de Magalhães

Description of a Glass Apparatus
For Making Mineral Waters

ISBN/EAN: 9783337337889

Printed in Europe, USA, Canada, Australia, Japan

Cover: Foto ©berggeist007 / pixelio.de

More available books at **www.hansebooks.com**

DESCRIPTION

OF A

GLASS APPARATUS,

FOR MAKING

MINERAL WATERS,

LIKE THOSE OF PYRMONT, SPA, SELTZER, &c.

In a few Minutes, and with a very little Expence:

TOGETHER WITH THE DESCRIPTION OF SOME

NEW EUDIOMETERS,

OR

INSTRUMENTS for afcertaining the WHOLSOMENESS of
RESPIRABLE AIR;

AND THE

METHOD OF USING THESE INSTRUMENTS:

IN A LETTER TO THE
REV. DR. PRIESTLEY, LL.D. F.R.S.

By J. H. DE MAGELLAN, F. R. S.

LONDON:
PRINTED FOR W. PARKER, No. 69, FLEET-STREET,
AND SOLD BY J. JOHNSON, No. 72, ST. PAUL'S
CHURCH-YARD, AND W. BROWN, THE CORNER OF
ESSEX-STREET, IN THE STRAND.
MDCCLXXVII.

ADVERTISEMENT

OF THE

EDITOR,

RELATING TO THE USE OF THE SIM-
PLE GLASS MACHINES FOR MAKING
MINERAL WATERS.

ALTHOUGH the method of using
the fimple glafs machines of Mr.
Parker, is minutely defcribed in the following Letter, neverthelefs it may be
more agreeable to find here diftinct refer-
ences to thofe articles, where it is particularly contained ; as the Author feem-
ingly could not help blending it with that
of his improved machines.

I. The defcription of the fimple Parker's machine is comprifed No. 5 and 6.

II. The procefs to make ufe of the
fame, is defcribed No. 9, 10 and 11.

III. The manner of carrying on the
production of *fixed air*, is indicated No.
14 and note (*d*).

IV.

IV. The method for reducing the procefs to a few minutes is defcribed No.
12, 13 ; fee note (*c*) and No. 15.

V. The manner of keeping thefe acidulous waters is fhewn No. 16.

VI. For making them fparkle like
Champaign wine, No. 17.

VII. For making them ferruginous or
chalybeate, No. 18.

VIII. As to the medical and œconomical
application of thefe waters, and *fixed air*,
fee No. 1, 3, 19 and 20.

N. B. *Both* glafs-machines, *for making arti--
ficial mineral waters, and the* Eudiometers *here defcribed, are made and fold at W. Parker's cut-glafs
manufactory, No.* 69, *Fleet ftreet, London.*

E R R A T A.

Add to the end of page 3, thefe words, *See Withers's
Obfervations on Chronic Weaknefs,* page 164, York, 1777,
in 8vo.

Page 39, laft line but one : *two or three* ; read *one or
two.*

Page 45, about the middle : $\frac{188}{200}$; read $\frac{198}{200}$.

THE

CONTENTS.

ON ARTIFICIAL MINERAL WATERS.

A

To

ON EUDIOMETERS.

(vii)

TO

REVEREND Dr. PRIESTLEY.

DEAR SIR,

I Do not know how better to employ the lei-
fure of thefe holydays I enjoy in your
neighbourhood, than in defcribing, according
to my promife (a), the two contrivances I have
mentioned in my laft letter; which, I hope,
will be ufeful to the public. It is with pleafure I have obferved a great agreement in
almoft all our philofophical ideas : but I am
very happy to find that we agree ftill more, in

(a) This refers to a former letter to Dr. Prieftley,
printed in the Appendix to his third volume of *Experiments and Obfervations on Air*, No. III, p. 376 of the
London edition.

looking with the greateſt indifference on any diſcovery, even the moſt ingenious, if no real advantage may accrue from it to mankind. Amongſt the many that you have made, and which are ſcattered in your Philoſophical Works, that of producing by art, at any time or place, with very little expence and trouble, Mineral Waters, like thoſe of Pyrmont, Spa, Saltzer, &c. whoſe virtues depend on their being impregnated with *fixed air* ; and that of finding out a general ſtandard or teſt for aſcertaining the greater or leſs ſalubrity of reſpirable air, in any place whatſoever, are, undoubtedly, the moſt beneficial. The ſucceſs with which the firſt of theſe two diſcoveries is employed, wherever it is known, and the very intereſting obſervations relating to the ſecond, made almoſt all over Italy by the Chevalier Landriani, with his Eudiometer, clearly evince the truth of my aſſertion.

2. As ſoon as your pamphlet, containing the method of making Pyrmont Water, fell into my hands, in the year 1772, I ſent abroad a great many copies of it, to different parts of Europe, where I have a literary correſpondence ; this having long been my cuſtom, whenever any uſeful diſcovery comes to my knowledge. I made then ſome change or improvement in your method, which rendered

4   the

the manner of conveying the fixed air to water somewhat eafier. This was added in a note to the French tranflation of it, made foon after at Paris, from a copy that I had fent to that great promoter of Natural Philofophy, Monfieur Trudaine de Montigny, mentioned page 268 of your fecond vol. on *Different Kinds of Air*. Sometime after, Mr. Blunt invented a machine, which rendered this operation ftill more eafy. One of this kind, made by himfelf at Mr. Nairne's, I fent to a very judicious lover of philofophical experiments at Turin, the Marquis de Rofignan ; now ambaffador of the King of Sardinia to the court of Berlin.

3. Another conftruction of a glafs machine for the fame purpofe, was publifhed by Dr. Nooth, in the 65th vol. of Philofophical Tranfactions. But this being very imperfect, was afterwards improved by Mr. Parker ; and you have given an account of it in the fecond volume of your work, above mentioned, page 298, and following. A very great number, above a thoufand, of thefe machines have been fent to different parts, even to the'Eaft Indies : and it is known that many perfons have been greatly benefited by the ufe of thefe artificial acidulous waters.

B 2   4. I

4. I found, however, not long ago, that the manner of conducting the procefs, as defcribed in the printed directions fent with thefe machines, was very inconvenient, on account of its flow operation ; it requiring four, and even fix or feven hours to get the water fully impregnated with fixed air. This I felt the more in November laft, on my being at his Serene Highnefs's the Duke of Arenberg, whom I have a right to call my Mecænas, on account of the many favours I have received from him. Knowing this generous Prince to be endowed with the beft difpofitions that any of his rank ever had, for encouraging and giving his protection to all improvements and difcoveries beneficial to mankind, this confideration prompted me to fend from London one of thefe improved machines to Bruffels for the ufe of his Highnefs ; and, on my trying it, on my arrival there, I felt, for the firft time, how difagreeable it was to wait fo long for the defired effect ; which could be foon compleated, if the firft method already mentioned, No. 2, was employed : for which reafon I always had made ufe of this laft, in preference to any other ; as it requires but few minutes to compleat the operation. I then confidered what could be done to avoid this. At laft I contrived the following apparatus, confifting of fome additional pieces, by which means the whole

whole operation is fo fhortened, as to take but few minutes: and, at the fame time, the quantity of the artificial water is increafed to double of that which is impregnated, at one procefs, in the fimple glafs-machine, improved by Mr. Parker.

A Description *of this* Apparatus.

5. Let A B C (fig. 1,) be one of the improved machines of Mr. Parker, ftanding upon a wooden difh *d e*, in order to avoid any water, if fpilled, from falling on the table. The middle veffel B has a neck, which is inferted into the mouth of the veffel A, being nicely ground air-tight to it. This lower neck of the middle veffel B has a ftopple V of glafs, compofed of two parts, both having holes, fufficient to let a good quantity of air pafs through them: between thefe two parts is left a fmall fpace, containing a plano-convex lens, which acts like a valve, in letting the air pafs from below upwards, and hindering its return into the veffel A.

6. The upper veffel C terminates below in a tube, marked 2, 1, (fig. 1,) which being crooked, hinders the immediate paffage of the bubbles of fixed air into the upper veffel C,

before

before they reach the furface of the water in the veffel B. The veffel C is alfo ground air-tight to the upper neck of the middle veffel B; and has a ftopple *w*, fitted to its upper mouth, which is perforated through the middle. This upper veffel C contains juft half the water that can be contained in the under one B; and the end (1,) of its crooked tube (2, 1,) goes no lower than the middle of the fame veffel B.

7. Befides thefe, I have added the two other veffels, G and H (fig. 2,) perfectly like thofe marked *B* and *C* (fig. 1,). The veffel H is furnifhed with a ftopper *i*, equally perforated as the other *w*, and contains half as much as the veffel G. Thefe veffels are fet upon the wooden ftand F : and the lower neck of the veffel G is not only furnifhed with a valve and ftopple, as already defcribed in the veffel B, but is fitted and ground air-tight to the neck of the fame veffel A. Each of the three vef-fels, A, B, and G, have an opening, *m*, *n*, and *l*, with ground ftopples, which are only open when occafion requires, as will be men-tioned hereafter.

8. I contrived the wooden ftand K, (fig. 3,) fo that a thick piece of glafs *x*, like a fmall tumbler, be cemented in the top, after it has

been ground air-tight to the under necks of the veffels B and G. The form of this ftand is eafily conceived by the engraved plate, fig. the 3d ; it being plane at the bottom, turns up in a kind of convexity z z towards its edge, and has a round moulding o o. Figure the 4th reprefents a wide glafs funnel q, which may enter into the upper mouth of the veffel A. Figure the 5th reprefents a fmall phial p, which ferves to meafure the quantity of the vitriolic acid to be made ufe of. Fig. the 6th reprefents a little trough of tin R, to meafure the pounded chalk or marble, that is to be employed in every procefs : and fig. the 7th reprefents a particular kind of ftopples, the ufe of which will be explained hereafter.

THE PROCESS.

9. Let fome dry chalk, as it comes out of the earth, that is to fay raw, and without being burned in the fire ; or rather white marble, which is much better for the purpofe (b), be reduced

---

(b) White marble being firft granulated, or pounded like coarfe fand, is much better for the purpofe, than the pounded chalk : becaufe the action of the diluted acid upon the marble lafts a very confiderable time ; and the the fupply of the *fixed air*, which is difengaged by this effer-

reduced to powder; and let fome oil of vitriol be at hand. The veffel *B*, together with *C*, (fig. 1,) muft be taken off from *A*, and put into the ftand *K* (fig. 3,). The veffels C and H being feparated from their under ones B and G; let thefe laft be filled with fpring, or any other drinking water, or even with diftilled water; and let them be joined again with the upper veffels C and H.

10. Let fome water be poured on the lower veffel A, fo as to cover the rifing part of its bottom: but if this appears too vague a direction, pour in fourteen or fixteen meafures of water, with the glafs *p* (fig. 5,): then fill the fame phial *p* with oil of vitriol, and pour it into the fame veffel A, along with the water.

Now let the glafs funnel *q* (fig. 4,) be put into the mouth of the veffel A; and filling the

effervefcence, is much more regular than otherwife. In general it continues to furnifh *fixed air* more than twenty-four hours. When no more air is produced, if I decant out of the veffel A, all the acid fluid, already faturated, and wafh off the thin, white fediment, I may employ again the remaining granulated marble, by adding to it frefh water, and a new quantity of vitriolic acid; which will then furnifh a further fupply of *fixed air*: and this may be repeated over again, until all the marble is diffolved; which will not be very foon.

trough

trough *R* (fig. 6,) with the pounded chalk or marble, let it be thrown into it. Take off the funnel *q*, which is used only to prevent the chalk from touching the inside of the mouth of this vessel : since otherwise it will stick so strongly to the neck of the vessel B, as not to allow the taking it off again without breaking. Then immediately place the two vessels B and C as they are, over the mouth of the vessel A ; and all the fixed air which is disengaged from the chalk or marble, by the force of the diluted acid, will pass up through the valve V into the vessel B. When this fixed air comes to the top of the vessel B, it will dislodge from thence as much water as its bulk : and this water, so dislodged, will go up, by the crooked tube 2, 1, into the vessel C.

11. Care must be taken not to shake the vessel A when the powdered chalk is poured in ; for otherwise a great and sudden effervescence will ensue, which will, perhaps, expel part of the contents. In such a case, it will be necessary to open the stopple *m*, in order to give vent to the effervescence for a moment ; otherwise the vessel A may happen to burst. Perhaps it will be necessary to throw away the contents, to wash the vessel with water (because the boiling matter will stick between the necks of these vessels, and will cement them to-

C   gether)

gether) and to begin the operation afresh. But if the powdered chalk is thrown in, without any confiderable shake of the machine, there will be but a small effervescence at the beginning. When this operates well, the vessel *C* will soon be filled with water, and the vessel *B* half filled with air ; which when done will be easily perceived, by the air going up in large bubbles by the crooked tube 1, 2 ; this will take place in about two or three minutes.

12. As soon as this is observed, take off the two vessels B and C, together as they are, from A; put them on the stand *K* (fig. 3,) and, taking the other two, G H, from the stand F, put them over the vessel A. Whilst the operation is going on in the vessels G H, hold, with the right hand, the vessels B C, which are now on the stand *K*, by the neck *z z*, inclining them a little sideways, and shake them very briskly, so that the water within B be very much agitated, presenting many fresh surfaces in contact with the fixed air, great part of which will be absorbed into the water ; as it will appear by the end of the crooked tube being considerably under the surface of the water in the vessel B.

13. It will suffice to shake the water in this manner during two or three minutes ; which done,

done, loofen the upper veffel C, fo that the remaining water may fall into the veffel B, and the unabforbed air may go out: then taking off thefe veffels from the ftand K, put them, joined together as they are, on the ftand F (c). By this time the veffel G will be half filled with fixed air, and the upper veffel H will be filled with the water, thrown up by it. Take then thefe veffels to the ftand K, and replace the others B C, on the mouth of the veffel A ; in order to be half filled again with fixed air, whilft the water in the veffels G H is brifkly fhaked in the fame manner as the others have been.

14. Whenever the effervefcence nearly ceafes in the veffel A, it will be revived again by giving it a gentle fhake, fo that fome part of the powdered chalk which is in a heap at the bottom of A, may be mixed with the diluted vitriolic acid, and difengage more fixed air. However, when it happens that the whole is exhaufted, and no more air rufhes up to the middle veffel from the lower one, either more powdered chalk muft be put in, or more oil of vitriol; or at laft more water, if neither of

(c) By this method, even the fimple glafs machines above mentioned No. 3, already defcribed No. 5 and 6, may be worked fo as to have the water fully impregnated in a few minutes, though with lefs advantage.

the

the two firſt produced the deſired effect. Theſe additions may be performed by letting them in, through the mouth of the veſſel A, whilſt the upper veſſels are changing. In this caſe uſe muſt always be made of the funnel *q*, in order to avoid the ſticking of the junctures above mentioned (*d*.).

15. When this operation has been repeated three or at moſt four times alternately, with each ſet of veſſels, by throwing out the remaining air which does not incorporate with the water, and putting in a freſh quantity of fixed air, the water contained in the two veſſels B and G, will be fully ſaturated ; and will be much more pleaſant to taſte than the natural Pyrmont or Seltzer's waters, which, beſides their fixed air (hardly the half of what this artificial water may abſorb) contain ſome diſagreeable ſaline taſte, which, it is known, does not contribute at all to their medicinal virtues ; but, on the contrary, it may be hurtful in ſome complicated caſes.

(*d*) Theſe directions muſt be reſpectively applied to the ſimple glaſs machines, mentioned No. 3, and deſcribed No. 5 and 6, when they are employed. If theſe veſſels be ſuffered to ſtand fix or ſeven hours, the water will be ſufficiently impregnated without agitation ; provided the ſupply of fixed air be copious.

16. Theſe

16. Thefe artificial waters will remain as limpid and as tranfparent as before, although there has been abforbed above, as much air as their own bulk. The whole procefs will hardly take above a quarter of an hour, by this method; and the quantity will be double of that which could be made in the fimple glafs-machine. The water may be taken out by the opening *l* or *n*, to be drank immediately; if not, it will be better to let it remain in the machine, where it has no communication with the external air: otherwife, the fixed air goes off by degrees, and it becomes vapid and flat; as it happens alfo to the natural acidulous waters. This artificial water may be kept a long time, in bottles well ftopped, placed with their mouths downwards.

17. In general they are fo fimilar to the natural acidulous waters, that they may be even made to fparkle like Champaign wine. Mr. Warltire has actually brought thefe waters to this ftate, by keeping the fixed air compreffed upon the furface of the water in the middle veffel; as appears by his letter printed in the Appendix to your third volume of *Experiments and Obfervations on Air,* page 366. The fame end will be obtained, if, inftead of the ftopples *w* and *i,* ufe is made of the folid one reprefented (fig. 7,) which has a kind of bafon
at

at the top, in order to hold fome additional weight. This ftopple muft be of a conical figure, and very loofe ; but fo well ground and fmooth in its contact, as to be air-tight by its preffure, which may be increafed by fome additional weights in its bafon. If the veffels are ftout enough, there is no danger of their burfting in the operation,

18. Thefe waters may alfo be rendered ferruginous (or chalybeate) very eafily, by putting, in the middle veffel, two or three balls of fine iron-binding wire : otherwife two or three ounces of fmall iron nails may be put in for the fame purpofe; becaufe they will diffolve iron as much as to be faturated with it in a few hours, according to the experiment of Mr. Lane. According to Sir John Pringle, there may be added from eight to ten drops of *tinctura martis cum fpiritu falis,* in order to refemble more nearly the genuine Pyrmont water,

19. No doubt thefe waters may be advantageoufly employed in many medical purpofes ; not only by diffolving in them the very falts which are found to be contained in many natural fprings, renowned for their different virtues, but even they may be applied as a vehicle to many draughts and internal medicines, which

which will be lefs naufeous to the patients,
and perhaps more agreeable to the ftomach,
giving to it a tonical ftrength.

20. I fhall conclude this fubject by obferving with you, that fixed air may be given to
wine, beer, cyder, and to almoft any liquor
whatfoever. Even when beer is become flat,
or dead as it is called, it may be revived by
employing the fame method : but the delicate,
though brifk, and agreeable flavour, or acidulous tafte, communicated by the fixed air, and
which is fo manifeft in water, will hardly be
perceived in wine or other liquors which have
much tafte of their own. I fhall now proceed
to the fecond object of this letter.

ON EUDIOMETERS.

21. The happy difcovery you have made
for the general benefit to mankind, and perhaps of almoft the whole animal creation of
this globe, by finding that *nitrous air is a true
teft of the purity of refpirable air*, which is ab-
folutely neceffary to life, and without which
it is prefently extinct, gives a moft ftriking
inftance of the blameable flownefs of mankind
to pay a proper attention to thofe objects, the
importance of which is infinitely fuperior to

I   that

that of the numerous trifling novelties which
fo often fpread with prodigious rapidity through
remote provinces, and even to the moft diftant
countries of the earth. Since the beginning
of the year 1772, in which you announced
this moft interefting and valuable difcovery,
in the 62d vol. of the *Philofophical Tranfacti-*
ons, no more than three or four philofophers
that I know of, have given any confiderable
degree of attention to fo important a fubject.

22. The Abbé Fontana and the Chevalier
Landriani, both of Italy, and already known
to the public by their feveral valuable pro-
ductions, were the firft, as it feems, who
availed themfelves of this difcovery (*e*). Both
propofed

(*e*) Mr. Volta, Profeffor of Natural Philofophy at Co-
mo in Italy, has made a difcovery, mentioned No. V. of
the Appendix to your third volume *On different Kinds of*
Air, which feems clofely connected with the prefent fubject. He difcovered that *inflammable air* is contained in
the mud of almoft all lakes, marfhes, and wet grounds of
Italy. He publifhed different letters on this fubject, of
which he was fo kind as to fend me a printed copy, after
part of this letter was printed. The experiments you have
made afterwards with me at Calne on this matter, fhow that
this air is lefs inflammable than what proceeds from the
folution of metals, with vitriolic acid : it burns with a
lambent flame, like the air produced by heat from char-
coal. This difcovery of profeffor Volta accounts very
well

propofed to the public one of the moft ufeful
kinds of inftruments that we can boaft of
among the numberlefs ones already employed
in philofophical refearches and experiments.
They gave to thefe inftruments different forms,
as appears by the printed defcriptions that each
of them has feparately publifhed : and the Chevalier Landriani has tranfmitted to England, as
a prefent to you, the very inftrument he had
made ufe of, to eftimate the refpective falubrity of the air in different parts of Italy, as
mentioned page 23 of the Preface to your
third volume *On different Kinds of Air.* This

well for the unhealthy habitations fuch marfhy grounds
generally afford to the human fpecies : and fhows the ne-
ceffity of examining, with care, by means of the Eudiometer, what places are fit for being inhabited. This is a
new and a very interefting requifite, never to be overlooked, before any building is erected, or the place for
any country feat is fixed upon. Such grounds or places
whofe atmofphere is loaded with phlogiftic miafma, are
the moft dangerous to animal life : becaufe the air of fuch
an atmofphere cannot be a good conductor or difcharger of
the fuperabundant phlogifton, of which the animal œco-
nomy requires to be unloaded : this being the aim intended by Nature in the function of refpiration, as you have
at laft difcovered, and incontrovertibly demonftrated, by
the moft decifive experiments, to be the cafe ; after fo many ineffectual attempts of the greateft philofophers of all
ages. This appears by fection V, page 55 and following,
of your third volume *On different Kinds of Air.* London
Edit. 1777, and by *Phil. Tranfact.* vol. 66, p. 226.

D   Eu-

Eudiometer is fmaller than that defcribed by him, and publifhed in the 6th volume of Rofier's Journal for the year 1775, though nearly of the fame form. It confifts of a glafs tube, ground to a cylindrical veffel, with two glafs cocks, and a fmall bafon, all fitted in a wooden frame. Quickfilver is there ufed inftead of water; and that part of it which replaces the bulk loft by the diminution of the two mixed airs, is conducted either through a kind of glafs fiphon, or through the capillary holes of a glafs funnel: fo that by its fall, the whole mixture of the two kinds of air is more readily made.

23. Dr. Falconer of Bath fent, fome time ago, to the Royal Society of London, a glafs tube, neatly divided; by means of which one may be enabled to know the quantity of diminution produced in a certain bulk of the mixture of *nitrous air* with another air, in order to judge of its falubrity, which you have fhewn to be in proportion to the *diminution fuffered in the fum of their original bulk, after they are mixed together.* This method is the neareft to your original one, or, rather, is the very fame you have ufed in the purfuit of this difcovery; as appears by your printed work on this fubject: and I think it to be the readieft of all, whenever no great nicety is required in obfervations

of

Ѧf this kind. There are, however, fo many circumftances neceffary in a good inftrument for fully anfwering fo great an object to its utmoft extent; that I fhould be deterred from offering to the public what I have as yet done on this fubject, was I not aware, that fome advantages always accrue to public good, by any new fteps towards perfection, how diftant foever we may ftill happen to be from its compleat attainment.

DESCRIPTION OF THE FIRST NEW EUDIOMETER.

24. Of the three Eudiometers I have contrived, which are reprefented fig. 8, 15 and 16 in the annexed plate, I think the laft is the eafieft in its application, and the moft exact in its refults. It is reprefented alfo (fig. 12, 14, and 17) in different pofitions, for the better underftanding of its application : and it confifts of the following parts : *viz.* a glafs tube *m n e d* fig. 16, about twelve or fifteen inches long, and of an equal diameter, with a ground glafs ftopple, *m* : a veffel, *c*, the neck of which is ground air-tight to the lower end *d* of the tube : and two equal phials *a* and *b*, whofe necks are alfo ground air-tight to the refpective mouths of the veffel *c*. Both thefe phials

contain

contain nearly as much as the whole tube *m n e d*. There is, moreover, a sliding brass ring, marked *z*, which slides in the tube *n d*, and may be made tight at pleasure by a finger-screw : and, lastly, a ruler, either of brass or of wood, represented fig. 11, which is divided into equal parts, and indicates the contents of both the phials *a* and *b*, when thrown into the tube, by the number of parts which is engraved or stamped about the middle of it. The two bent pieces of brass *z t* serve to hold it easily by the side of the tube *n d* fig. 14 and 17, keeping it close to its neck *n* by the notch *i*.

25. Experiments with these Eudiometers, which are easily constructed, may be made either with water or with quicksilver ; with this difference, that when the last is made use of, the Eudiometers (particularly the third, represented fig: 8, which seems the fitest for being used with quicksilver) will be more convenient if made of a still smaller size. Mercury, however, is a fluid that, I think, never ought to be used preferably to water, in the inside of Eudiometers ; because it suffers a sensible action from the contact with nitrous air, as yourself have observed : and this must have an influence on the result of the experiments. Water, on the contrary, seems less liable to mistakes, although it imbibes some part of the *ni-*

trous

trous air. In fact this effect only takes place in a long time, or with much agitation : and, after duly weighing the question on both sides, I should think water may be generally used, without the fear of any sensible error. The weight and the dearness of quickfilver, are, likewife, two other confiderations to give the preference to water in thefe experiments.

THE PROCESS.

26. In the firft place there muft be either a trough, as reprefented fig. 17 ; or at leaft a common tub, nearly filled up with water, un-lefs the tall glafs receiver, of which I fhall fpeak No. 34, be at hand. I take out the ftopple *m* (fig. 16,) fill the Eudiometer entirely with water, keeping it in the pofition repre-fented fig. 16 and 17. I then fhut it with the ftopple *m*, without leaving any bubble of air in the infide; and put the lower part *c* under the furface of the water in the tub (fig. 17) in an erect pofition as it is therein feen. I take the phial *a*, filled with water; and, keeping its mouth downwards under the furface of the water, I fill it with that air, the falubrity of which I want to afcertain *(f)*. This is done
either

(f) The cafe I am fpeaking of, is when I have a bottle
of

either by putting the phial *a* on the shelf *n̈ a*
of the tub fig. 17, and throwing the air into thé
glass funnel *t*, which is there cemented to the
shelf ; or by holding in the left hand the same
phial *a*, together with the glass funnel *B*
(which is reprefented fig. 18, and has no pipe
at all) applied to the mouth of the phial,
whilft I pour the air with my right hand into
it. But left the heat of my hand should produce any confiderable expanfion in this air, I
generally ufe, in hot weather, the wooden tongs,
reprefented fig. 21, with two bent wires *x x*;

of air, which has been taken at any diftant place, and fent
for trial. If a glafs bottle, with a ground glafs ftopple,
is filled with water or with mercury, and emptied in the
place whofe atmofpherical air is intended for being examin-
ed, it will, of courfe, be filled with that air: and, being
clofely fhut with the glafs ftopple, may be carried to any
diftant place for trial. By this means the atmofpherical
air of any part of a country may be fent to any diftant one,
in order to afcertain its comparative falubrity: and many
ufeful inquiries and difcoveries may be made hereafter
on this fubject, with great eafe, and at very fmall expence.

But if I only want to try the air of the room, where I
have the Eudiometer, I then only pour out of the phial *a*
the water it contains: I find that, however, after fome trials
with *nitrous air*, the atmofphere about me is loaded with
phlogiftic miafma: and for that reafon I always empty the
phial *a* out of the window of the room, in order to have
nearly the fame kind of air in all the experiments.

in

in order to hold the glaſs funnel *z* cloſe to the mouth of the phials ; unleſs they are made with a ſolid lump at their bottoms, as repreſented in the plate. *See note (g).*

27. The phial *a* being filled with that air, the ſalubrity of which I am to examine, I put it into the mouth of the veſſel *c*, making it rather tight,

(g) There are ſome niceties to be obſerved in order to fill up, exactly, any phial intended to ſerve as a meaſure of air, of which I muſt give an account in this place. The eaſieſt method to ſucceed is the following : · Let a glaſs funnel *t* (fig. 17) be cemented under the hole *s* of the ſhelf *s o* in the trough. In this caſe I hold the phial *a* filled with water, with its mouth downwards over the hole *s* of the funnel *t*: I throw the air into the funnel ; and, when the phial is filled with air, I take it ſideways, rubbing its mouth along the ſurface of the ſhelf, ſo that the redundant air, adhering to the mouth of the phial, be got off : and I put it into the mouth of the Eudiometer belonging to it. But as the heat of the hand muſt expand the air contained in the phial, which of courſe will then contain leſs air than its real meaſure in the temperature of the ſurrounding water, I handle the phial with a kind of pliers or tongs of wood, repreſented fig. 21, till the neck enters into the proper place of the veſſel *c*, where I ſecure it with the other hand : and, laying aſide the wooden tongs, I make it properly tight. But if the phials have a ſolid knob at their bottoms, as repreſented in the plate, it will then be enough to handle them by it only : ſince the heat of the hand can not be communicated in ſo ſhort time to the air in the inſide.

If

tight, which muſt be done with ſome care ;
for if the phials *a* and *b* are not tight enough
to the reſpective mouths of the veſſel *c*,
they will flip out, when turned downwards,
and of courſe will be broken : and, if they
are too tight, the veſſel *c* will be eaſily cracked,
and become unfit for uſe, The better to
avoid theſe accidents, and to judge of the
proper degree of tightneſs, let the necks of
the phials *a b*, and of the veſſel *c*, as well as
the glaſs ſtopple *m*, be always rubbed with
tallow, previouſly to every experiment. When
I have done with the phial *a*, I take the other
phial *b*, filled with water : by the ſame method

If I have not the convenience of a trough, prepared
with a ſhelf, and its fixed funnel, as above mentioned,
an aſſiſtant holds the funnel under the water in a common
tub, whilſt I fill up the phial with air : and I take care to
hold the phial in ſuch a manner that the end of the funnel
be out of the inſide of the phial at the laſt moment, that
the air may ruſh out after it is totally filled : otherwiſe
that part of the phial, occupied by the end of the funnel,
will not be totally filled with the air.

Even without any aſſiſtant, but with a little care, a per-
ſon may hold both the phial and the funnel in the left
hand, whilſt he throws the air into it with the other hand ;
as I have myſelf frequently done in experiments of this
kind : and when I make uſe of the wooden tongs, I add to
it the two bent pieces of wire *x x* (fig. 21) by means of
which the funnel is kept cloſe to the mouth of the phial.

I throw

I throw into it as much *nitrous air* as to be perfectly filled up with it: and I then replace this
phial *b* in the other mouth of the veffel *c* (*b*).

28. I take afterwards the Eudiometer with
my left hand, holding it near the lower part

d;

(*b*) No pains or trouble ought to be fpared, in order to
obtain, at any time, a *nitrous air* perfectly alike in its
contractive power, when mixed with common air.

In order to come the neareft to this, I take a phial *D*
(fig. 19,) like thofe you have defcribed in the fecond volume
of your work *On different Kinds of Air :* to the mouth of
which is ground air-tight the crooked tube *n z* in the fhape
of an S. I fill the half of this phial with thin brafs wire,
the thicknefs of which is equal to $\frac{1}{70}$ of an Englifh inch,
nicely cut by a pin-maker to this length. I fill the three
quarters of the phial with common water ; and the re-
mainder with ftrong *nitrous acid,* which I have always
taken of the beft fort, at the Apothecary's Hall in London. I put the crooked tube *n z* to the phial : and, as foon
as the effervefcence caufes the liquor to rife to the end *z* of
the tube, I pafs it under water into the mouth of the bottle
Z (fig. 20,) which is filled with water, and inverted with
its meuth downwards upon the hole of the fhelf *n o;*
which appears covered with water within the trough, or
pan, (fig. 17).

This figure reprefents the moft commodious fhape a
trough muft have for any experiments on different kinds
of air. It is made with ftrait boards of elm-wood *one
inch* thick. The infide dimenfions are 25 inches long,
13½ wide, and 11 deep, Englifh meafure. The two end
boards, *c d* and *e f,* are fitted into a groove cut in the other

E   three

d, over the furface of the water in the trough, to avoid breaking any of the phials, if it chan=
ces

three boards ; this is daubed with thick white painting, as a cement, to keep well the water in : and the whole is faftened with nails from the outfide. The fhelf *w a n s* is eight inches wide, and two inches thick. It has three holes of three tenths of an inch diameter, with as many fe-parate cavities underneath, fo as to ferve like fo many fun-nels. The figure, however, reprefents a glafs funnel *t* ce-mented to the middle hole *n :* which is equally convenient. This fhelf is fupported by four metallic hooks *V w z x*, which may be raifed or lowered at pleafure, by the wooden wedges there reprefented.

When the bottle *F* is entirely filled by the *nitrous* air, I fhut it up with its ftopple *x* (fig. 20,) which I pafs under the furface of the water, to avoid any communication with the external air : and I pufh this bottle under the fhelf, where I let it remain for a quarter of an hour, to acquire the fame temperature of the furrounding water : and the fame I always obferve with the bottle, containing that atmofpherical air which I defire to try, before I put it into the phial *b*.

I muft acknowledge, however, that, notwithftanding thefe precautions, I cannot fay that all the refults of my expe-riments, even when made upon the fame atmofpherical air, have as yet agreed fo exactly as I flattered myfelf they would. Perhaps there was fome difference in the ftrength of the *nitrous* air, the denfity of which I thought might eafily be brought to a fettled ftandard, to be determined by means of a glafs hydrometer. Perhaps there was fome other little variety in the circumftances of the experiments, the influence of which I was not aware of. But let it be as it may : I very willingly leave this problem to be re-folved

ces to fall ; and, with my right hand, I turn the veſſel *c* upwards, ſo that the two phials may be downwards, as repreſented fig. 14. By this operation the two kinds of air come up to

ſolved by abler chemiſts than I can pretend to be : and I heartily wiſh they may ſucceed better than I have done : for, without being aſſured of getting every where a cer‑ tain *ſtandard nitrous air*, by which the ſame atmoſpherical air be equally affected, we cannot draw with certainty any general deciſive concluſions, from Eudiometrical experi‑ ments made in diſtant times or places.

Before I leave this ſubject, I cannot help mentioning two ſtriking circumſtances relating to *nitrous air*. The firſt is the great quantity produced by the action of *nitrous acid* on many metals ; which may ſtill be carried to a greater extent, if helped by bringing the flame of a candle to the phial, which contains the ſolution, when it ſeems to be nearly done with. The ſecond is the antiſeptic power of *nitrous air* to preſerve animal matters from corruption. A beef-ſtake, almoſt entirely putrid, and with an inſupport‑ able ſtench, being put into a jar of *nitrous air*, in leſs than two days was perfectly reſtored, and very eatable when dreſſed. A pigeon was very well preſerved above ſix weeks by the ſame treatment ; and, when roaſted, was found ſo good as to be eat without any diſlike. Two other pigeons were kept in it full ſix months without corruption : they were ſtill very firm and of a good colour ; but the fleſh had loſt all its flavour, and was far from being eatable when dreſſed. But the *nitrous air* for theſe œconomical purpoſes, which may be of a great reſource at ſea, as well as at home, muſt be made out of *nitrous acid* with iron, or other metal leſs exceptionable than braſs or copper, the effluvia of which are pernicious to animals.

x, from the phials *a b*; and there they mix together in the beſt poſſible manner; the particles of each having a large room to come into contact with each other; ſince the foremoſt ones do not detain thoſe which are behind, as it happens when this mixture is made in a narrow veſſel. This being done, I immediately dip the Eudiometer in the water of the trough (fig. 17,) leaving the mouth of the inſtrument above its ſurface; ſo that no more water may enter into it than what it h..d at firſt. I then obſerve with attention the moment when the mixture *x* (fig. 14,) of the two kinds of air comes to its greateſt diminution, after which its bulk will begin to increaſe again. In order to catch this moment with certainty, I ſlide down the braſs ring *z* of the inſtrument, as the ſurface of the water in the tube falls. This point of the greateſt diminution will be eaſily perceived, by obſerving when that inſide ſurface is ſtationary : which will happen in a few minutes, if the *nitrous air* has a proper ſtrength *(i)*.

29. As

(i) The bulk of the mixed air will decreaſe to a certain degree, within a few minutes, according to the ſtrength of *nitrous air*. Afterwards it will begin to expand again : but this it will do to a very ſhort limit, much below its former bulk. This is a phenomenon which, I think, I have obſerved the firſt on theſe experiments ; having

29. As foon as the diminution of the two kinds of air appears to be ftationary, I fill up the whole tube of the Eudiometer with water: I fhut it up with the ftopple *m* ; and incline the top of the inftrument forwards, till the air comes from *x* (fig. 14,) up to the top *n* of the tube. I then keep the lower part of the inftrument dipped in the water ; take off the glafs veffel *c* with the two phials *a b*, and rife or lower the tube of the Eudiometer, fo as to fee the furface of the water, in the infide, even with that in the outfide ; which I mark by fliding to it the brafs ring *z*. Otherwife I apply the ruler (fig. 11, without making any ufe now of the brafs ring) to the fide of the Eudiometer, whilft it is immerfed in the water of the trough : and there I fee the true dimenfion of the remaining bulk of the two kinds of air, already diminifhed. Perhaps the beft method for this obfervation would be to allow time enough that the mixed air may take its fettled bulk : but this requires fometimes twenty-four hours time. I leave, however, the choice of thefe two methods to the obferv-

ing made a very great number of them with nice Eudiometers, of the kind I am now defcribing. It certainly deferves the attention of Philofophers : and, although I have communicated it to fome of my acquaintance, none have as yet, in my humble opinion, given a fatisfactory folution of this phenomenon.

er,

er, who may ufe both if he pleafes, provided he keeps diftinctly the refult of each method in his account of the experiment.

30. The number marked about the middle of this ruler (fig. 11,) as for inftance, * *=96, means that the contents of both phials *a* and *b* are equal to ninety-fix divifions of the ruler, when put into the tube of that Eudiometer : that is to fay, they are equal to a folid cylinder, as thick as the infide of the glafs tube, and whofe length is ninety-fix divifions of the ruler, which has been divided into tenths of an Englifh inch.

31. Now if, for inftance, this remaining bulk of mixed air correfponds to the 56th divifion of the ruler, it fhews that, out of 96 parts, only 40 (=96—56) have been loft or contracted : and, in this cafe, the wholfomenefs of that air, which I call A, will be $\frac{40}{96}$. If another equal quantity of different air, which I fhall call *B*, had alfo been tried by the fame Eudiometer, and its refiduum was equal to 60 parts of the fame ruler, the refpective falubrity of the air *B* will then be to that of the air *A*, as 36 (=96—60) to 40.

32. But if the air *B* had been tried by another Eudiometer, whofe proportional dimenfi-

ons,

3

ons, marked about the middle of its ruler, were $*\ * = 108$, then the refpective falubrity of thefe two kinds of air A and B, would be in the compound ratio of $\frac{36}{108}$ to $\frac{40}{96}$

$$= \frac{36 \times 96 \text{ to } 40 \times 108}{108 \times 96} = 3456 \text{ to } 4320 = 54 \text{ to}$$

67, 5 : that is to fay, the wholfomenefs of the air B would be to that of the air A, as 54 to $67\frac{1}{2}$ (k).

33. Nearly the fame refults would be found, if the ruler (fig. 11,) was applied to the fide of the

(k) It is fuppofed that the infide of the tube is of an uniform diameter ; but it often happens, that there are fome varieties in different parts of its whole length. When they are not very confiderable, we may neglect their influence in the refult of thefe Eudiometrical experiments ; but, when the contrary happens, it will be very eafy to make a proper allowance for them in the calculation. It is for this reafon that I have always ordered that the contents of one fingle phial be marked alfo upon the fcale of each Eudiometer, as well as the contents of both phials ; for inftance as in this manner :

$$*\ * = 96$$
$$* = 47$$

which means, firft, that the contents of both phials a and b are equal to a cylinder, whofe diameter is the fame as that of the infide bore of the tube $n\ d$ (fig. 16,) and whofe height is equal to 96 equal divifions of the ruler : fecondly, that the contents of a fingle phial are equal to 47 divifions in the upper part of the fame tube $m\ n\ d$; and
of

the Eudiometer, as foon as the inclofed mixture of air came to its utmoft diminution, as
mentioned No. 28: becaufe as much water
muft fall in the tube *n d*, as correfponds to
the diminution fuffered by the two mixed airs
in *x*. But there are fome varieties, which arife
from the different preffure of the column of
water, which preffes more or lefs upon the air
at *x* (fig. 14,) as it is longer or fhorter : and
thefe varieties ought not to be overlooked in
nice experiments : they are avoided by the
procefs already defcribed, No. 29 ; and may
otherwife be prevented by the method of
which I will fpeak at the end of No. 39.

34. Whenever I have at hand a tall glafs
receiver, like that reprefented fig. 14, the
whole procefs is then more eafily performed :
for in this cafe I dip the Eudiometer, inverted
as it appears fig. 12, into the water contained
in the veffel *V S q l* : I then put the two kinds
of air into the phials *a* and *b* as above faid, No.
26 and 27 : I turn the inftrument upright, as
reprefented fig. 14 ; and finifh the procefs, as I
have already defcribed.

of courfe, to 49 divifions ($=96-47$) of its lower part.
By this difference it appears that the tube of fuch Eudiometer is wider in the top than at the bottom, by $\frac{2}{56}$ of the
whole.

35. I muſt, however, warn the operator that, unleſs every trial, and even almoſt every part of the proceſs, be made in the ſame temperature ; or, at leaſt, unleſs the varieties ariſing from this cauſe be accounted for no reliance can be had on the reſult of ſuch experiments : it being well known, that air is apt to increaſe or diminiſh very conſiderably in its bulk, by the influence of heat and cold. It is for this reaſon that I conſtantly keep a good thermometer K, which hangs by the wire y r, and is immerſed in the water of the glaſs veſſel fig. 14, or in the trough fig. 17, whenever I make any of theſe experiments. For the ſame reaſon I take care to leave the Eudiometer and the veſſels of air, immerſed in water time enough, as above mentioned, to get the ſame temperature : and I make uſe of the wooden tongs mentioned in note (g), whenever I handle the phials c b filled with air, chiefly if they have not the ſolid lump at their bottoms, as repreſented in the plate; unleſs I feel the heat of my hands to be the ſame as that of the water, in the trough, I make uſe of.

DESCRIPTION OF THE SECOND NEW EUDIOMETER.

36. The Eudiometer, repreſented fig. 15, conſiſts of a glaſs tube t c, two or three feet

. F . long.

long, and of an uniform diameter : the end c is bent forwards ; and the other end t is wide open, as a funnel, unlefs a feparate one is made ufe of : this tube is faftened, by two loops, to the brafs fcale $c\ w\ t\ V$. There is a glafs phial n, the neck V of which is ground air-tight to the end t of the tube; and contains only half of the whole infide capacity of the divided tube $c\ t$. It has, at the other end c, a large round phial $a\ b\ c$, containing three or four times the bulk of the phial n : its neck is alfo ground air-tight to the mouth c of the tube. The brafs fcale $c\ w\ t\ V$ is divided into 128 equal parts: this being a number that can be divided to unity in a fubduplicate ratio without fraction, by continual bifections ; on which account it is one of the numbers the late famous Mr. Bird had adopted for his dividing mathematical inftruments with the utmoft accuracy. Thefe numbers are fet out in the fcale from t towards c. The contents or capacity of the tube till the number 128 is the double of the capacity of the phial n. Befides this there is a tin veffel $x\ s\ d\ t\ r\ o$ (fig. 15*), which may ferve as a packing cafe for the whole inftrument, and its neceffary appendages ; and alfo as a trough, when experiments are made; it being then filled with water. Both the glafs tube reprefented fig. 22, and the glafs ftopple m (fig. 15*), belong to
this

this Eudiometer; and both are fitted in, air-tight, to its mouth V.

THE PROCESS.

37. Let the instrument be immersed under the water $z z$ of the tin vessel fig. 15*: and let the phial n, filled with water, be put in the inside socket $e e d$ of the tin vessel. Let it be filled with *nitrous air*, as above directed at the end of No. 27: and let this quantity of air be thrown into the phial $a b c$ (as directed No. 26 and No. 27), which I fix a little tight to the mouth c of the Eudiometer. I afterwards fill the same phial n with the air I want to try: and, raising the end c of the instrument, I put it into its mouth V: when this is done, I set the instrument upright, as represented fig. 15, hanging it on the hook w; and, as soon as this last air goes up to the phial $a b c$, I take off the phial n, that the diminution of the two mixed airs may be supplied from the water in the tin vessel; which must be the case, as the mouth V of the Eudiometer is then under the surface of the water.

38. I then put to the lower end V of the Eudiometer, the bent tube fig. 22, to which is fitted the brass ring K, and is filled with

water

water. It is by obferving the furface of the water in this fmall tube (which then forms a true
fiphon with the tube of the inftrument) and
by means of the brafs ring K, that I can dif-
tinguifh the ftationary ftate of the diminifhing
bulk of the two mixed airs, above mentioned
at the end of No. 28 : which being perceived,
I take off the fmall tube g h from the Eudiometer, and lay down, for fome minutes, the
whole inftrument, in an horizontal pofition,
under the water of the tin veffel : I fhut up
the mouth V with the glafs ftopple m; and,
reverfing the inftrument, I hang it up by the
end V, on the hook w. By this pofition
the whole diminifhed air of the veffel a b c goes
up to the top, where its real bulk is fhewn by
the number of the fcale, facing the infide furface of water. This number being deducted
from 128, gives the comparative wholfomenefs of the air already tried, without any further calculation.

39. But this procefs will be ftill eafier, when
the laft diminution of the two mixed kinds of
air, mentioned No. 29, is only required in the
obfervation : becaufe no ufe will be then made
of the fyphon (fig. 22). In fuch a cafe the inftrument is left hanging on the hook w for 48
hours : after which it is laid down under the
water of the trough (fig. 15*), in an horizon-

tal

tal pofition, for 8 or 12 minutes, in order to acquire the fame temperature of the water : the mouth V is then fhut up with the ftopple m; the inftrument is hung by the end V in a contrary pofition, and the laft real bulk of the good mixed air will be then fhown by the number of the brafs fcale anfwering to the infide furface of the water. This number being fubtracted from 128, will give the comparative falubrity of the air employed in the trial, without any further calculation. I need not fay that all the circumftances already mentioned for the better obtaining exact refults in thefe experiments, are to be carefully obferved, when this fecond, or the third Eudiometers are ufed: but chiefly that circumftance, mentioned No. 35, ought never to be omitted. The thermometer is to be kept dipped in the water of the tin veffel; and the Eudiometer muft be kept there immerfed fome minutes, as I have faid juft now, before it is raifed for the laft time, to read off the quantity of the total diminution of the mixed air. As to the other circumftance, mentioned No. 33, it has been rendered unneceffary by laying this Eudiometer in an horizontal pofition before the glafs ftopple was put in. The fame method muft be applied to the third new Eudiometer I am going to defcribe; and even the firft Eudiometer, already defcribed, may be treated in the fame manner : for if it be

laid

laid down in an horizontal pofition under the water in the tub, before it be fhut up with the ftopple, as directed No. 29, there] will be no variation produced by the expanfion of the air in the infide : becaufe the proper quantity of water is then fhut up within the glafs veffel *c* of the inftrument: fo that raifing it up, as it is, together with the veffel *c*, and its phials *a b* (fig. 14), the weight of the column of water will prefs totally upon them, without expanding the inclofed air, or caufing any variation beyond the trifling one which may proceed from the natural elafticity of the fides of the glafs tube and veffels.

40. I muft, however, acknowledge that, the long way through which the air paffes, in going at firft to the large phial *a b c* in this fecond Eudiometer, muft leave fome doubt whether it has not then fuffered fome fenfible change in its quality, before it is mixed with the *nitrous air* ; fince, as you have obferved, the air that has been long agitated in water, changes for the better from its bad qualities : and this objection muft be ftill greater in the ufe of the third Eudiometer. It is on this account that I have mentioned the firft Eudiometer, as the leaft exceptionable of all that we know till the prefent ; and perhaps the nature of the thing is not capable of a further perfection.

Indeed

Indeed that inftrument, I mean my firft Eudiometer, has not only the advantage of offering a very fmall way through the water to the
two kinds of air, on their going to mix at *x* in
the veffel *c* (fig. 14), but they are kept feparated till that moment, in the two refpective
phials *a* and *b*, without any other contact with
the water, but only in the narrow diameter of
the necks of thefe phials.

DESCRIPTION OF THE THIRD NEW EUDIOMETER.

41. This third Eudiometer is the neareft to
your original one: and, was it not for the confideration I have mentioned in the preceding
number (which, perhaps, will not weigh too
much with fome philofophers) and few other
circumftances which are obvious, I fhould not
doubt to pronounce this third inftrument to be
the beft of all the three, as I have advanced in
my laft letter to you of the 30th of November
laft. This, however, I gladly fubmit to your
fuperior judgment.

42. This third Eudiometer confifts of a
ftrait glafs tube *e n* (fig. 8), of an uniform diameter, about two or three feet long, with a large
ball *s*, and a glafs ftopple *m*, fitted air-tight to
the

the mouth *n*, which ought to be wide open, as a funnel, unless a separate one is made use of. There is also a small siphon (fig. 23) with a brass ring *x*: a small phial *z* (fig. 9), the contents of which may be received in the third part of the ball *s*: and, when put into the glass tube *n s*, must take there no more than the half of its length. Lastly, this instrument has a ruler (fig. 13), which is divided and stamped like that other already described at the end of No. 24; and a glass funnel, which is ground to the mouth *n* of the instrument, when this is not wide open, as already said.

THE PROCESS.

43. The use of this instrument is easily understood by what I have already said of the two preceding ones.

First, it is filled with water, and set in a vertical position, with the mouth *n* under the surface of the water in a tub, or in a trough, (fig. 17).

Secondly, the phial *z* (fig. 9) is filled, as above, with *nitrous air*; and thrown into the tube by means of the glass funnel *y* (fig. 10), which is ground to the mouth *n* of the Eudio-

meter,

meter ; unlefs it be wide enough not to be in need of any funnel.

Thirdly, the fame phial z is again filled with the air to be tried; and thrown into the fame.

Fourthly, the fiphon (fig. 23) is added immediately to the mouth n of the Eudiometer, under the furface of the water; fome of which is to be poured into it.

Fifthly, the ftationary moment of the greateft diminution of the mixed air at s, is watched by means of the ring x, as mentioned No. 28 and 38.

Sixthly, when that moment arrives, the fiphon $K l$ (fig. 23) is taken off; the Eudiometer is laid for fome minutes under the water, in an horizontal pofition, or nearly fo, but in fuch a manner that no part of the inclofed air may get out ; the mouth n is fhut up with the glafs ftopple m, and the inftrument is inverted with the mouth n upwards.

Laftly, the fpace occupied by the refiduum of the diminifhed air, is meafured by applying to its fide the divided ruler, or fcale (fig. 13), and the refult is eftimated after the manner already explained No. 31 and 32.

G

44.

44. Whenever I want only to know the laſt diminution of the mixed air, mentioned No. 39, the proceſs then becomes eaſier, as no uſe is made of the ſiphon (fig. 23). The method of conducting the proceſs in ſuch a caſe being reſpectively the ſame as that already deſcribed No. 39, it is unneceſſary to deſcribe it here again. The ſame precautions I have ſpoken of, No. 35 and 39, muſt be obſerved when this Eudiometer is made uſe of, in order to form a true judgment concerning thoſe places, where people will be able to live without danger of hurting their conſtitutions, by breathing and being continually ſurrounded by noxious air; which they have not yet been able to diſtinguiſh from the moſt wholſome, except by a long and too late experience.

45. The Eudiometers already deſcribed are the fiteſt inſtruments for philoſophical experiments, on the bulk of air and other fluids, when mixed together; and even when mixed with ſome ſolid ſubſtances, which can be introduced into the lower veſſel c of the firſt of the three Eudiometers. It will be better, however, to have them made purpoſely for ſuch objects, with a tube two or three times longer than I have indicated above. Whenever dephlogiſticated air is to be tried by theſe inſtruments, proper care is to be taken to obſerve the preciſe point of its full ſaturation,

3

which

which is that of its greateſt diminution by the addition of *nitrous air.*

46. In order to make this experiment with great accuracy, let a narrow glaſs tube of an uniform diameter (fig. 24), be provided : let one of the two phials *a* or *b* (fig. 16) filled with quickſilver, be thrown into it, and the tube cut exactly to that ſize, ſo as to contain neither more nor leſs. Let its whole length be divided into ſome number of equal parts, by which number the value marked on the ruler (fig. 11), of this Eudiometer, can be divided without any fraction : for inſtance, the number ✳ ✳ = 108, marked in the ruler, means, that the contents of the two phials *a* and *b*, of which I ſpoke No. 32, are equal to a cylinder of 108 diviſions long, as thoſe of the ruler : and, of courſe, it ſhews that a ſingle phial *a* or *b* contains but 54 of theſe parts. In this caſe this tube (fig. 24) may be divided either into 27 parts, each containing two of the ruler ; or into 54, into 108, &c.

N. B. If the top of the tube is not very flat in the inſide, it will be more exact to divide the weight of the quickſilver in two parts ; to put one of them into the tube ; to mark the ſpace occupied by it ; to divide the part of it, which was empty, into half the number

in-

intended for this tube, and afterwards to divide the other half into fimilar equal parts, as
the firft half, carrying them towards the clofed
end.

47. If the dephlogifticated air is very pure,
it will require almoft double the quantity of
nitrous air to be completely faturated. In order to do this without exceeding the neceffary
quantity, I throw into the tube $n\,d$ (fig. 17)
a fecond meafure b or a of *nitrous air*, after I
have brought the procefs to the moment mentioned No. 29: in this cafe the whole volume
or bulk of the dephlogifticated and nitrous air
will be 162 [=108+54 :] I obferve where the
furface of the infide water in the tube ftops,
and I mark it by the fliding brafs ring z. I
then fill up the divided tube (fig. 24) with
nitrous air : I throw a fmall quantity into the
Eudiometer tube $n\,d$; and, if it becomes of a
redifh colour, the inclofed air will diminifh :
I then pufh up the ring z, and, by this means,
I go on throwing in the nitrous air, by little
and little, till I fee that the whole diminifhes
no more ; which fhews me that it is fully faturated.

48. Let us fuppofe, for example, that the
tube (fig. 24) was divided only into 27 equal
parts; and that the faturation of the dephlogifticated

gifticated air was compleated at the eighth divifion of it : this fhews that 19 parts [27—8
=19], equal to 38 of thofe marked in the ruler, have been thrown into the Eudiometer;
that is to fay, that the whole bulk of both
kinds of air is equal to 200 [= 162 + 38]
meafures, as thofe marked by the ruler (fig.
11,) already explained No. 30. Now if the
remaining quantity of air within the Eudiometrical tube is only equal to two meafures or
numbers of the ruler, it is clear that fuch dephlogifticated air is ninety-nine times of an
hundred $\left[\dfrac{200-2}{200} = \dfrac{188}{200} = \dfrac{99}{100}\right]$ pure air;
fince its bulk is reduced, by the combination
of *nitrous air*, to the $\frac{1}{100}$ of the whole.

49. It is but three days ago (*l*) that you
fhewed me fuch a wonderful kind of air, as I
have exemplified in the preceding number.
This air you have produced before my eyes,
from a folution of *quickfilver* and *nitrous acid*,

(*l*) This additional article to the prefent letter was wrote
on the 16th of September, 1777 ; although the greateft
part of it has been written many months before, and the
firft twenty numbers were already printed : but fome circumftances, the knowledge of which cannot intereft the
public, have hindered the publication of the whole till
this prefent time.

made

made many months before, and then diftilled in a long but narrow glafs retort, with a fand-heat. This is, perhaps, indeed, an extraordinary phenomenon, and feems to bring us a little nearer to the door of the fecret laboratory of Nature in the formation of air.

50. I cannot fay, but fo pure a *dephlogifticated air* may ftill be produced by this procefs; that its whole bulk may be reduced to nothing by a proper combination with *nitrous air*. If fo, what fhall we then be able to think of a fluid fubftance, which is coercible in a glafs veffel, to which above the double quantity of another fubftance [$\frac{1+6}{3+4} = 2{,}7$], likewife coercible in a glafs veffel, being added; both thefe fubftances to appearance wholly vanifh!

51. This phenomenon certainly deferves the attention of philofophers: and I gladly leave to them the examination of it. I muft only add, for their information, that the *nitrous acid* is the thing chiefly concerned in its production: when this admirable fubftance acts on certain kinds of bodies, as *quickfilver* in the prefent cafe, its folution produces that *elaftic*, but *coercible* fluid, which we call *nitrous air*: the refiduum, after a long while, being properly urged by fire, gives at laft the other *elaftic*, but likewife *coercible* fluid, which we call

call *dephlogisticated air* ? and the combination
of both, nearly in the above proportion, produces the wonderful phenomenon I have spo-
ken of.

52. I shall say no more on this matter; and
leave it very willingly to be considered and un-
ravelled by abler philosophers than I can pretend to be : and conclude the subject of this
letter by assuring you, that I shall be very
happy, if the things here treated of should
deserve your approbation : and still more so,
if they produce the desired effect I aim at,
—the general good of mankind. I am, with
the utmost regard and sincere friendship,

  My dear Sir,

   Your most obedient and

    Affectionate servant,

Bowood Park, J. H. de MAGELLAN.
January 3, 1777.

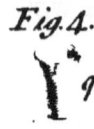

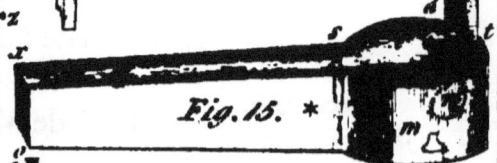

Fig.1.

Fig.2.

R Fig.6.

Fig.4.

Fig.5.

Fig.3.

Fig.7.

Fig.9.

Fig.11.

Fig.10.

Fig.8.

Fig.13.

Fig.13.

Fig.15. *

Fig.16.

Fig.19.

Fig.18.

Fig.20.

Fig.24.

Fig.21.

Fig.23. Fig.22.

Fig.17.

J. Lodge Sculp.